JUDICIAL CONTROL OF GOVERN

CONCEPT OF LAW FOR A CAF

Lectures on the Common Law

Editors: Professor B.S. Markesinis and Judge J.H.M. Willems

Volume 1

J.G. Collier: Judicial Control of Government Action
R.W.M. Dias: Concept of Law for a Caring Society

LECTURES ON THE COMMON LAW

Volume 1

Judicial Control of Government Action

J.G. Collier

Concept of Law for a Caring Society

R.W.M. Dias

Kluwer Law and Taxation Publishers
Deventer · Antwerp
London · Frankfurt · Boston · New York

Distribution in the U.S.A. and Canada
Kluwer Law and Taxation Publishers
101 Philip Drive
Norwell, MA 02061
U.S.A.

Cover design: B. Betzema

D/1988/2644/78

ISBN 90 6544 378 9

FOREWORD

The year 1978 marked the thirtieth anniversary of Professor C.J. Hamson's 'Summer Course for foreign lawyers' and of his untiring efforts not only to introduce the civil lawyer to the mysteries of the Common Law but also to bring closer together lawyers from both sides of the Channel. The same year saw the beginning of the 'Cambrigde-Tilburg Law Lectures' which developed indirectly from the 'Summer Course' and which seek to achieve similar aims by different methods.

The success of the lectures far exceeded the expectations of its most ardent supporters, and the programme received a great boost when, six years after it was launched, it acquired the support of the University of Leiden. As we now approach the tenth anniversary of the lectures, developments in the Netherlands – including the recent establishment of the Institute of Anglo-American Law in Leiden – underscore the continuing interest in the Common Law, so it was decided that the time was ripe to rename the lectures. By calling them Lectures on the Common Law the organisers wish to stress their determination to broaden what was originally an enterprise between individuals of two law faculties into a more ambitious scheme that will involve more Universities on both sides of the Channel.

The immediate objects are to achieve closer links between Common lawyer and Civil lawyer, to encourage the further systematic teaching of the Common Law and to produce a series of lectures, two of which will be published annually in the hope that they may be of interest to a wider public.

Professor Hamson died in November 1987. His 'Cambridge Summer Course' was the *fons et origo* of the Anglo-Dutch co-operation. He would, we suspect, be happy to hear that the programme is going from strength to strength and that more foreign law students and lawyers are wishing to continue their legal studies in Britain, in part at least as a result of the excellent work done by our visiting lecturers. We have every reason to believe that the success of the first decade will continue and grow in the second.

B.S. Markesinis,
Denning Professor of Comparative Law in the University of London,
Professor of Anglo-American Law in the University of Leiden

J.H.M. Willems,
Judge in the Court of Appeal of Amsterdam,
Reader of English Law in the University of Tilburg

Table of Contents

Judicial Control of Government Action 1
Central Governmental Powers: The Prerogative 4
Control of the prerogative by the courts: the orthodox view 6
The same: a new view 8
Can and should the prerogative be subject to judicial control? 14

Concept of Law for a Caring Society 23
Just allocation of benefits and burdens 33
Curbing abuse of power 37
Curbing abuse of liberty 42
Just decision of disputes 47
Adaption to change 48
Concept of law for a Caring Society 48

JUDICIAL CONTROL OF GOVERNMENT ACTION

J.G. Collier M.A., LL.B.

Fellow and Tutor of Trinity Hall,
Lecturer in Law in the University of Cambridge

JUDICIAL CONTROL OF GOVERNMENT ACTION

One of the most striking developments in English law during the last twenty-five years has been the increased preparedness of the courts to intervene in order to control the exercise of governmental action, whether the governmental activity in question is that of central government or that of local government authorities. I wish here to investigate one or two aspects of this phenomenon in connection with central government and to inquire whether this much increased judicial activity is in all respects advisable or satisfactory.

When I was a student of law in the mid-nineteen fifties, it did not take me long to master the principles and rules of administrative law. The courts and the writers of text books, who took their cue from the great Oxford jurist, A.V. Dicey, pointed to the absence in England, as compared with France, of any separate system of administrative courts or law distinct form the ordinary courts and law of the land. Indeed, there was some self-congratulation over this. Administrative law filled the last few chapters of text-books on constitutional law, the rest of which were devoted to explaining the constitutional structures of the United Kingdom (itself no easy task when there exists no written constitution). This state of affairs existed until the last years of that decade, which saw the publication of the late S.A. de Smith's 'Judicial Review of Administrative Action' and a slim volume on Administrative Law by Professor John Griffith and the late Professor Harry Street.

Moreover, the courts seemed to collude with this view. During the nineteen thirties, under the influence of such judges as the Lord Chief Justice Hewart[1], and after the Second World War, with the coming to power of the Labour Government of 1945 armed with a mandate form the electorate to indulge in state activity in areas which had hitherto been regarded as matters of purely private and commercial nature and to establish what became known as the welfare state, the courts showed themselves to be in spirit non-interventionist. Thus, for example, they went out of their way to deny the relevance of the rules of natural justice in cases where decisions of ministers and officials were concerned. (The rules of natural justice, evolved by the courts, require absence of bias on the part of one who has to pronounce upon the rights of individuals (nemo judex in causa sua) and that a person whose rights are affected is entitled to a fair hearing of his case (audi alteram partem)). Decisions such as that of the House of Lords in Franklin v. Minister of Town and Country Planning[2] and of the Judicial Committee of the Privy Council in Makkuda Ali v. Jayaratne[3] led the distinguished authority on constitutional and administrative law, H.W.R. Wade, to write of 'the twilight of natural justice'[4].

But this attitude began to change in the early-nineteen sixties. Starting with the decision of the House of Lords in Ridge v. Baldwin[5] in 1963 which, though it was not concerned with central government, can be taken as the case which restored natural justice to what many regarded as its natural place, things proceeded apace as the courts, through the highest tribunal in the land, especially in Anisminic v. Foreign Compensation Commission (1967)[6] and Padfiels v. Secretary of State for Agriculture (1969)[7] began indeed to strain to find new ways of intervening and controlling the exercise by tribunals and ministers of the statutory powers which until that time had seemed to be virtually uncontrollable through the courts. The process was described by Lord Diplock in giving the judgment of the House of Lords in O'Reilly v. Mackman[8] and was said by him to have resulted in our having a system of administrative law. The process continues today.

Partly by reason of these developments individuals and corporations and local government authorities are nowadays more conscious of their rights and more knowledgeable of the existence of legal means of redress against the government and its departments. Moreover, fairly recent reforms in the procedure for obtaining judicial review of administrative action and in the increased availability of appropriate remedies have led to a situation where the courts are being almost overwhelmed by applications for judicial review even though there exists now, as a subdivision of the Queens' Bench Division of the High Court, a judicial body which can be called an administrative court[9].

Before I turn to my central thesis, it is important to observe that in reviewing the exercise of governmental powers the English courts do not concern themselves with deciding on the merits, that is the rights and wrongs of the course of action adopted by the authority in question. That is said to be the task of that authority on whom the power has been conferred, be it a government minister, a chief officer of police, a local government authority or whoever. When an authority is given a power to do something it alone has that power and it is given the discretion as to how to exercise it. Thus, if a chief constable of a police force has the power to dismiss a police officer, the courts will not interfere with his decision simply because, though the chief constable thought the officer was unsuitable to continue to hold that office, the judge thinks he is suitable to do so[10]. The courts are only willing to intervene if they think that the power and the concomitant discretion were not properly exercised because, for example, the authority in question did not act in accordance with the rules of natural justice or that it did not itself exercise the discretion but allowed someone else to do so[11], or that the decision arrived at is on the face of it so clearly absurd or unreasonable that it suggests that the authority took into account irrelevant considerations or omitted to take into account relevant ones. (This last has come to be referred to as the 'Wednesbury' test, after the decision in which it was first clearly enunciated by the Court of Appeal)[12]. Most important in the context of the present discussion, the courts will always be astute to enquire whether the action which it is sought to impugn was intra vires of the authority, that is, not outside or beyond the power which it has purported to exercise (ultra vires) as it will have been if the authority can be shown to have acted in orde to achieve an object or in pursuance of a purpose for whose attainment it was not granted the power.

These principles of administrative law apply to both central and local government, and indeed to every body to which Parliament has entrusted a power. Two distinctions must however be drawn between organs of central government and of local government, and I shall hereafter be relatively unconcerned with the latter. First, central government is not a creation of Parliament; it was not set up by a statute. But local authorities are; they exist entirely at the will of Parliament[13], as witnesses the recent abolition of the Greater London Council and the six Metropolitan Counties. Second, all local government powers[14] are conferred on local authorities by statute, but only some of the powers of central government. Some important powers of the Crown (the legal term for central government) are recognised by the courts as being part of the ordinary common law of the land. These are known as 'prerogative' powers.

Central Governmental Powers: The Prerogative

The legal powers of the Crown derive from two distinct sources, then. These are (i) the royal prerogative and (ii) statutes, and these must be carefully distinguished from one another.

The prerogative powers look rather curious nowadays. They are the powers which the Crown (at that time synonymous with the monarch himself) was in the Middle Ages recognised by the courts as possessing, and which have not been taken away from it since then. They are legal powers which the Crown alone possesses and no other person[15]. It is not always very easy to state with confidence whether a power is a prerogative power or whether it is a power which anyone else may possess, in which case it is not a prerogative power. But it is clear, for example, that the Crown's right to spend money which Parliament has granted it without stating any specific purpose for its use is not a prerogative right, but until the seventeenth century its power to raise money by taxation almost certainly was a prerogative power[16].

It is convenient for the purpose of exposition to divide the prerogative powers between (a) the so-called 'personal' prerogatives, which the Queen herself exercises, almost always on the advice of the Prime Minister or her ministers generally, such as the appointment of the Prime Minister, the creation of peers and the dissolution of Parliament, and (b) the so-called 'political' prerogatives. These are in fact exercised by the Prime Minister or ministers themselves acting on behalf of the Queen. They include the appointment and removal of ministers and civil servants, powers in connection with the administration of justice such as pardoning convicted criminals or at one time reprieving those who had been sentenced to death, powers concerned with the national defence such as the command and disposition of the armed forces and national security in general and powers relating to the conduct of foreign affairs, for example, the powers of declaring war and of entering into treaties with foreign states.

As I have already said, the prerogative is the residue of the powers which it possessed in early times and which the courts of law recognised. It forms part of the common law of the land. Like the rest of English law, prerogative is at the mercy of Parliament which can legislate so as to restrict it, regulate its use or abrogate it alltogether. Parliament can do this directly and expressly as it did in the Bill of Rights of 1689 which itself marked the triumph of Parliament over the King and of statute over the prerogative. This enactment provided, inter alia, that the power of the Crown (or which the Crown claimed) to raise taxes without the consent of Parliament was illegal, and made unlawful the raising of a standing army in time of peace without Parliamentary authority. Parliament may do this not only directly but also by implication only, as where it legislates over the area covered by the prerogative power but in a manner inconsistent with the latter. In Attorney-General v. De Keyser's Royal Hotel[17] in 1920 the House of Lords held that the Crown could not take private property for its use in a time of national emergency by relying on a prerogative power to do so, because Parliament had legislated for such taking of property on payment of compensation to the owner. The Crown had to pay compensation to the dispossessed owner of the hotel. Lastly, the prerogative may be limited indirectly as was held by the Court of Appeal in Laker Airways v. Secretary of State[18] where Parliament had passed legislation dealing with the licensing of airlines to fly on the north Atlantic routes. It was held that the Secretary could not refuse such a licence to Laker by reliance upon the prerogative relating to treaty relations with other countries.

Control of the prerogative by the courts: the orthodox view

Since, as we have seen, the prerogative is part of the common law of the land the courts can control and regulate it in a certain sense. The traditional or orthodox view which has often been expressed in decided cases, is that the courts have the right to define the limits of the prerogative power in issue and even to say whether it does or does not in fact exist. However, once they have done this, they have no competence to go further, as they have in respect of statutory powers, to review the Crown's exercise of the prerogative in the way described earlier. A modern case in which the court took care to define the power in a sense contrary to what the Crown contended for is Burmah Oil v. Lord Advocate[19] in 1965. In that case, the plaintiff company sought compensation from the Crown for loss caused to it by the Crown, acting through the armed forces when its oil wells in Burmah were destroyed by the British army in its retreat from the invading Japanese in order to prevent them from falling into the enemy's hands in 1942. The Crown argued that it was entitled to do this without incurring any liability to pay compensation, just as if the damage had been caused during an actual battle. The House of Lords held, however, that the case was not the same as battle damage and (relying on Hugo de Groot amongst others as authority) that the Crown could only legally do what it had done if it paid compensation to the property owner. (This somewhat unreal distinction and, in my view, the almost perverse decision of the House gave the plaintiff company a short-lived and empty victory for Parliament immediately enacted the War Damage (no. 2) Act 1965 which provided with retrospective effect that no compensation should be paid in such cases).

The case is somewhat misunderstood and is sometimes claimed as support for the proposition that the courts can review the actual exercise of the prerogative. But this is clearly not so; the court was not called upon to determine whether the Crown's decision to blow up the oil-wells was ultra vires or an improper or unreasonable exercise of a power (this would have been absurd). It merely stated exactly what the prerogative power in fact was.

There exists plenty of judicial authority which denies any right of the court to control the exercise of the prerogative powers. In China Navigation Co v. Attorney-General[20] (1932) this was said of the Crown's power to dispose of the armed forces. In Chandler v. Director of Public Prosecutions[21] (1974) the House of Lords refused in effect to question the Crown's judgment with respect to matters affecting national security. In Hanratty v. Lord Butler[22] the court declined to question the exercise of the prerogative of mercy by the Home Secretary. Perhaps one of the most significant recent cases in this sense is Blackburn v. Attorney-General[23], a case wherein Mr. Blackburn, an indefatigable litigant where public affairs are concerned, attempted to move the courts to restrain the Crown form becoming a party to the Rome Treaty and so a member of the European Economic Community. The Court of Appeal, headed by Lord Denning, M.R. emphatically denied that the court could assist Mr. Blackburn; the right to enter into treaties is a prerogative right and the manner of its exercise can only be called into question elsewhere than in the courts.

Where then can the exercise of the prerogative be questioned? The answer, in the accepted model of the British constitution is, in Parliament. This is because it has been made clear since 1689 that the Crown in its governmental capacity is constitutionally responsible for governing the country subject to the political control by Parliament. The powers of government which were left with the Crown as part of its prerogative after the Revolution Settlement embodied in the Bill of Rights were those which it was realised were required by it in order to carry on the running of the country. As to those powers which can only be exercised or are excercised by the monarch personally apart, the practice or convention evolved that the powers of government should be exercised by Ministers appointed by the Crown who should be members of Parliament and themselves be answerable to Parliament for their conduct of government. If Parliament were to be dissatisfied with the conduct of government it could, and sometimes did, overturn it. This is what is meant by Parliamentary or Cabinet Government and by Ministerial Responsibility, two of the conventions of the British constitution. Indeed, many of the conventions of the constitution which are, speaking in general terms, practices which are not dictated by rules of strict law but are regarded as binding rules of constitutional conduct, were evolved in order to regulate the exercise of the prerogative.

The same: a new view

But within the last ten years there has appeared another, novel doctrine to the effect that the courts in some cases at least review the exercise of the prerogative, though I suggest that this has not yet in fact been put into effect. It is said that the court may review governmental action taken under the prerogative if the issue comes before the court in a 'justiciable form', whatever that may mean. The origin of this seems to have been a remark, a single word even, made by Lord Devlin in Chandler v Director of Public Prosecutions when he said 'The courts will not review the proper exercise of discretionary power (Query: how does one discover whether the exercise is proper until one has reviewed it?) but they will intervene to correct excess or abuse' [24]. Later, in Laker Airways v Secretary of State[25] Lord Denning said that the court could review the exercise of the prerogative, in marked contrast to his judgment in Blackburn's[26] case. But the authority which he cited for the proposition was Attorney General v De Keyser's Royal Hotel[27] whose actual ratio decidendi I have tried to expound earlier, and the case is not at all in point. The other members of the Court of Appeal proceeded to their decision along orthodox lines.

Another case which is put forward in support of the new view is R v Criminal Injuries Compensation Board, ex parte Lain[28]. In this case the Queen's Bench Divisional Court held that it possessed the power to review the conduct of the Criminal Injuries Compensation Board in arriving at its determination of a case before it. The Board, whose function is to make awards out of public funds to persons who have suffered injury as a result of criminal acts, was created by the Government under the prerogative. It was not created by Act of Parliament as are most courts and tribunals of a similar character. The decision of the Divisional Court is said to demonstrate that the courts are able to review the exercise of the prerogative and indeed the House of Lords or some members of it viewed the case in that light in the more recent case of Council of Civil Service Unions v The Minister for the Civil Service (1984)[29] (known as the G.C.H.Q. case) which I will consider shortly. Indeed Lord Roskill described ex parte Lain as a 'landmark decision' and Lord Scarman, with remarkable exaggeration, described that case as 'comparable in its generation with the Case of Proclamations[30] (1611) in its'.

Quite apart from the extravagance of the language employed by their Lordships, this analysis of ex parte Lain is, I submit, wrong. The decision merely illustrates the elementary proposition that the courts will always insist that any tribunal, not merely the courts of law, which is created in order to determine a person's culpability or rights or entitlement should behave like a court and hear a case properly. This is so whatever the source of the tribunal's authority, be it statute or prerogative, or even where it is not a public tribunal at all but a 'domestic' tribunal created essentially by contract or agreement, like the rule of an association. For example, the courts have from time to time been concerned with disciplinary or investigatory committees which have powers over the members or officials of an association, of the Football Association [31], the Greyhound Racing Board[32], Trade Unions[33] and Political Parties[34]. The decision in ex parte Lain would support the proposition contended for if what had been in issue was not the conduct of the Board but the propriety of the Crown's decision to create the Board in the first place, for it was the creation of the Board, not the latter's decision, which was the exercise of the prerogative.

Before I proceed to discuss the G.C.H.Q. case, it is important to identify those considerations which have led to the new view being formulated and advocated. It seems to me that there are two basic reasons.

The first, which is based on a political evaluation, is that the constitutional organ which, according to constitutional theory, is supposed to control the exercise of governmental authority, Parliament, does not do it very well. This is because instead of Parliament controlling government, government nowadays controls Parliament. There was, perhaps, a 'golden age' of the constitution, which lasted for a few years or decades in the mid-nineteenth century. Its 'chronicle' is Walter Bagehot's English Constitution, published in 1867. This age spanned the period after 1832, when the First Great Reform Act widened the electorate beyond the very small number of persons theretofore qualified to vote in Parliamentary elections (it was further widened in 1867 by the Second Reform Act). It is possible to some extent to say of the period that Government became controlled by the electorate through the House of Commons. But after 1867 when the electorate became very large in numbers, it became necessary for the political parties form whom the governments were formed to organise the electorate in their support and to harness its votes so as to elect the leaders of one or other of the two parties to form a government. (There have only been really two main parties, Conservative and Liberal, the latter succeeded in this century by Labour, with the exception of the years between 1918 and 1931). The main tast of members of Parliament became that of maintaining their party's government in power and not, save in exceptional circumstances, to control them by the ultimate treat of turning it out of office. Apart form the fall of Chamberlain's Government in May 1940 (and he was not actually defeated in the vote in the House of Commons) a government has only resigned after a vote in the House on two occasions, when Labour Governments resigned in 1924 and 1979, but on both of those the government had only the support of a minority of members of the House of Commons at the time anyway.

So, if Parliament is not doing the job it is supposed to do, who is more appropriate to take the task than the courts?[35]

The second, legal, reason is this. As we have seen, the prerogative powers were those which Parliament chose to leave in the hands of the Crown in order to enable it to carry on the government of the country. Until the mid-nineteenth century the business of government was very limited. It was expected only to make sure that the country and its empire were defended against possible deprevations by foreign countries and, at home, to preserve the safety and integrity of the persons and property of those who were owners of property from the lower orders in society. But things changed and now are very different. Today, any government, whether its philosophy causes it to like it or not, is involved and is expected to be involved, in almost every aspect of life from defence to education, the environment, transport, trade and industry and more or less to look after us from the cradle to the grave. In order to carry out all these functions the Crown needed more numerous and more specific powers and these have been provided by Parliament in the form of statutory powers. Therefore today the most important activities of government in connection with the day to day running of the country are derived from Acts of Parliament. From a practical every day perspective the prerogative powers are relatively insignificant.

Now, as we have seen, the exercise of statutory powers, whether these are conferred on the Crown or its ministers or anyone, is subject to the control of the courts. It seems odd, almost paradoxical, therefore, that action taken under the prerogative should not be subject to judicial review.

Some examples of relatively recent cases, drawn from the very many notable ones of recent years, which illustrate the judicial control of the exercise of statutory powers, may be given here. In Padfield v Secretary of State for Agriculture[36] the House of Lords intervened to compel the Minister properly to consider representations made to him in respect of a scheme for the marketing of milk, in his exercise of a statutory power. In Congreve v Home Secretary (1976)[37] the Court of Appeal condemned the misuse by the Minister of his statutory power to issue and revoke television licences. In the Laker case, as we have seen, that court also disallowed an action of the Minister in the purported exercise of a statutory power to license airlines to fly on certain air routes. In the celebrated Tameside[38] case in 1977 the House of Lords condemned as unreasonable the conduct of the Secretary of State for Education in exercising a power under the Education Act regarding the provision of education by a local education authority in a case in which that Minister had previously decided to be unreasonable the conduct of a newly elected district council in seeking to delay the introduction of a system of comprehensive education in place of the existing selective system. More recently, the Minister responsible for roads and transport has fallen foul of the courts on a remarkable number of occasions.

The House of Lords, or some of its members, explicitly referred to the reason now under consideration in the G.C.H.Q.[39] case. This litigation arose out of the decision of the Government to ban trade unions among the Crown's employees at G.C.H.Q., the Government's security communications centre at Cheltenham and its subsequent decision to terminate the employment of such civil servants as did not agree to renounce their union membership. The ban on trade union membership was a change in the employees' conditions of service and was made under an order which empowered the Minister for the Civil Service (who is also Prime Minister) to give instructions for controlling the conduct of the home civil service. The question before the House was whether, since the instruction was given without prior consultation with the civil servants concerned or the trade unions which they had previously been allowed to join, was void because of the lack of observance of the proper procedure.

The problem was that the instruction was given in exercising a prerogative power. The House of Lords discarced the settled law. Lords Diplock, Roskill and Scarman held that the exercise of prerogative powers is in principle just as subject to judicial review as is that of statutory powers. Lords Fraser and Brightman were a little less expansive, holding only that the exercise by someone of authority delegated to him or her under the authority of the prerogative was reviewable.

But one suspects that the House of Lords, having made what at first sight looks to have been a great constitutional breakthrough, did not really have the courage of their convictions, or noted how difficult it would be to carry out this newly assumed tast. For they limit the situations in which review would be possible to 'justiciable' prerogative powers. This limitation will be discussed later. But Lord Roskill expressly said that foreign affairs powers, the defence of the country, the granting of honours, the dissolution of Parliament and the appointment of ministers as well as others which he did not specify, were not fit for judicial scrutiny. In fact the only power that was held to be subject to review was the regulation of the terms and conditions of employment of civil servants, the one in issue in the case itself. This seems to be a somewhat tame conclusion.

In any event, it was held that judicial review was not available even in this case where the action taken by the Minister had to be taken for reasons of national security. It seems, as was once said on the first night of a new play, that there is 'less in this than meets the eye'.

Can and should the prerogative be subject to judicial control?

One thing should be clear from the foregoing account. The cases and judicial statements which support the thesis that the courts can review the exercise of prerogative powers in the same way as they do that of statutory powers are extremely few, limited, and somewhat tentative. Further, it may be that, in spite of brave words, the judges are somewhat hesitant about claiming a very active role in the matter. This provokes one to ask two questions: (1) Can the courts really do what is claimed? and (2) Should they do so? As regards (1) I submit that there exist very real difficulties, which do not stand in the way of the review of the exercise of statutory powers.

First, as the House of Lords realised in the G.C.H.Q. case, there is the preliminary problem presented by the idea that the courts can intervene if the issue is presented to them in a 'justiciable form'. This really means that it must be suitable for adjudication by a court. The circularity is obvious and like all circles only gets us back where we started from. The phrase 'justiciable form' or 'justiciable dispute' has been judicially employed before, but usually in the negative rather than a positive sense, meaning that the court should decline to decide the case on its merits because it is not justiciable because, for example, that would involve the court in passing upon the rights and obligations of foreign sovereign states[40], or that for the court to entertain the case might prejudice international negotiations for the settlement of a dispute[41]. One suspects that the words really only signify cases which the court does or does not wish to deal with and that their juridical significance is nil.

But there are more fundamental legal problems about it. As we have seen, the purpose of judicial review is simply to determine whether a decision has been arrived at properly or improperly. The usual ground on which it is sought to impugn such a decision is that the authority in exercising the power in question abused it, by not using it for a proper (i.e. authorised) purpose, for example, by using a statutory power to construct a public lavatory not really in order to achieve that objective (which it did indeed accomplish) but so as to construct a pedestrian subway under the street, which it had no power to do, and has merely built the lavatory underground as an adjunct to the subway and as an attempted justification for building it[42].

A statutory power is one granted by Parliament and set out in a document (the Act of Parliament) which can be examined in order to see just what power was bestowed on the authority and to see whether it expressly or by a process of interpretation reveals the purpose for which it was granted. But a prerogative power was not granted to the Crown by anyone; it was merely left by Parliament and the courts in the Crown's hands. It is not contained or set out in any document, so there is nothing at which the courts can look so as to discover what purpose is meant to be attained by exercising a prerogative power. It is not therefore possible to say whether the decision made or the action taken by the Crown or on its behalf was improperly made or taken and therefore ultra vires.

There is a counter-argument to this. The reason why the prerogative powers as they now exist were left to (and not taken away from) the Crown was, as I have suggested, that they were the powers it needed for the purpose of the government of the country. It can be inferred that there is a single purpose for which the prerogative powers are to be employed; that it should advance government, or to use a favoured term, the prerogative should be exercised 'for the public good'. Now this argument is insidious. If the courts are to review the exercise of the prerogative they will have to determine not only, as is at present the case with statutory powers, that the decision or action should advance the proper purpose and not be used for a collateral purpose, but they will also have to decide what is for the 'public good'.

This is a most elusive concept, to say the least, and is a farcy from the erection of a public toilet or municipal bathhouse or maintaining the flow of traffic or even the advancement of education, all of which are clear or tolerably so. In fact, of course, a judgement about what is for the purpose of government or the public good cannot be anything other than a political judgment. That is why in

British constitutional theory such judgments are left to politicians, the government and the parliament elected by the public to represent it to supervise the conduct of government in the light of its judgment of what the public good requires. What special qualifications do the judges have for the task of making such judgements? They are learned in the law, not trained in politics or administration[43].

Moreover, however feebly Parliament controls Government, I myself am at least involved in the election of a member of Parliament and ultimately the selection of a Goverment; it is my view of the public good which I can express when voting. Even if I backed a loser, at least I had the choice. I have no choice in the selection of the judges, and do not think them to be better qualified to represent my view of the public good than a doctor, a clergyman, a professional footballer or anyone else.

I may add that not everyone has been wildly pleased at the judges' keenness to intervene in the exercise of statutory powers. In Congreve's case, for instance, the result was that some televison owners and viewers effectively got licences to see their programmes more cheaply than did the vast majority; the latter might well feel annoyed. The issue in the Tameside case was an acutely political one. It was a skirmish in the war between those (though not all) on the political left who support the principle of comprehensive education and those, mainly on the right, who do not. It was a newly-elected Conservative district council which was elected in support of the latter faction who won the case. The losers scarcely thought the outcome satisfactory and thought the decision to have been politically motivated, however much it might be protested that it was not. A third example, which did not involve central government was Bromley Council v Greater London Council[44] in 1983 (the Fares Fair case) in which the courts, up to and including the House of Lords, declared ultra vires and void the policy adopted by the newly-elected very leftwing Labour Greater London Council of reducing fares on London Transport by subsidising them at the expense of the ratepayers. As law, the result was no doubt impeccable but the opportunity given to the party of the left to complain of a politically inspired judgment against them and of a judgment which ignored the express wishes of those who had elected the council, was unfortunate.

There is, nowadays, a certain amount of criticism of the English Judiciary on account of their social background, education and training and these are said to be such as to prevent the judges having other than right-wing political attitudes[45]. Be this as it may (though the criticism is not itself beyond criticism), it has always been the boast of the judiciary that they are above politics. This is, I believe, true on the whole. But in modern times, partly because of the political nature of the formally legal questions which are thrust upon them, and because of a desire of some litigants to fight what are really political battles through the courts, it is hard to see how the judges can altogether avoid becoming to some extent 'politicised'. I believe that the efforts of some judges and courts in administrative law may tend to further this and the matters I have been discussing show the possible dangers and difficulties which may ensue.

Fundamentally, whether one regards all this as a good or a bad thing depends upon whether one wants a politicised judiciary or not. I hope that what I have said shows that I do not, and the reasons for my view.

1. Before his appointment as Lord Chief Justice, Hewart had been Attorney General, and so the government's chief legal adviser. Before his appointment he had written a work, The New Despotism, published in 1927. It was highly critical of the increase in government powers. His attitude to these changed considerably on and after his attainment of government office.

2. [1948] A.C. 87.

3. [1951] A.C. 66.

4. [1949] 10 Cambridge Law Journal, 216.

5. [1964] A.C. 40.

6. [1969] 2 A.C. 147.

7. [1968] A.C. 997.

8. [1984] 2 A.C. 237.

9. This is a result of a revised order 53 of the Rules of the Supreme Court, which regulates the procedure on applications for judicial review. See also Supreme Court Act 1981, s. 31.

10. See, for example, Chief Constable of North Wales v Evans [1982] All E.R. 141.

11. See Barnard v National Dock Labour Board [1953] 2 Q.B. 18.

12. Associated Provincial Picture Houses v Wednesbury Corporation [1948] 1 K.B. 23.

13. This was made clear in Bromley Council v Greater London Council [1983] A.C. 768.

14. Local Government Act 1985.

15. Blackstone defined Prerogative as 'that special pre-eminence which the King hath, over and above all other persons, and out of the ordinary course of the common law in right of his regal dignity'. (Commentaries on the Laws of England, I, 239).

16. This was held to be the law in Hampden's case (The Case of Ship Money) (1637)3 St.Tr. 825.

17. [1920] A.C. 508.

18. [1977] Q.B. 643.

19. [1965] A.C. 75.

20. [1932] 2 K.B. 197.

21. [1964] A.C. 763.

22. [1971] 115 Solicitors' Journal 386.

23. [1971] 2 All E.R. 1380.

24. [1964] A.C. 763 at 810.

25. [1977] Q.B. 643.

26. [1971] 2 All E.R. 1380.

27. [1920] A.C. 508.

28. [1967] 2 Q.B. 864.

29. [1984] 3 All E.R. 935.

30. The Case of Proclamations (1611) 12 Co. Rep. 74 is the authority for the rule that the Crown cannot legislate without the authority of Parliament and is therefore one of the central cases in constitutional law.

31. Enderby Town v Football Association [1971] Ch. 591.

32. Pett v Greyhound Racing Association (1970] 1 Q.B. 46.

33. See Breen v A.E.U. [1971] 2 Q.B. 175.

34. John v Rees [1970] Ch. 345.

35. Lord Denning has been quite explicit about this.

36. [1968] A.C. 997.

37. [1976] Q.B. 629.

38. [1977] A.C. 1014.

39. [1984] 3 All E.R. 935.

40. Buttes Gas and Oil Co. v Hammer (1981] 3 All E.R. 616.

41. Lord Denning M.R. in Hesperides Hotels v Aegean Turkish Holidays [1978] 2 Q.B. 205. The other two members of the Court of Appeal and, later the House of Lords, gave judgment on a totally different ground.

42. Westminster Corporation v London and North Western Railway Co. [1905] A.C. 426. The proper purpose was held to be the predominant one, therefore the exercise of the power was lawful.

43. Some high court judges, though scarcely any of the present ones, have been members of Parliament and very occasionally of governments in their earlier careers, though usually they have been Attorney- or Solicitor-General.

44. [1983] A.C. 768.

45. See, for example, J.A.G. Griffith, The Politics of the Judiciary (3rd. ed 1985).

CONCEPT OF LAW FOR A CARING SOCIETY

R.W.M. Dias M.A., LL.B.

Fellow of Magdalene College,
Lecturer in Law in the University of Cambridge

CONCEPT OF LAW FOR A CARING SOCIETY

This is not an exposition of law, but about law, that is to say, it is an essay in legal theory or "jurisprudence". A minor hope is that the paper will convey some idea of what this word means in Britain since it bears different meanings in different countries. No preliminary definition of so a morphous a term is likely to be of any use; only a broad, general idea is possible at any time, which will have to be gathered from this paper as a whole -if the reader is patient enough to bear with it to the end[1]. In short, the scope of "jurisprudence" has to be shown rather than explained. A preliminary caution is that the paper is not going to be the effort of a jurist looking at the social implications of some aspects of law, but rather that of a jurist looking at a social phenomenon and considering its juristic implications, a distinction which, hopefully, will become apparent[2]. The chosen phenomenon is our contemporary fetish, the Caring Society.

What, now, do we understand by this? It is one in which we expect the state to assume obligations to conduct an increasing number of activities, which were formerly not its concern, but were left to individuals. For example, in the past health, education, care of children, the sick and the elderly, trade and industry and so on were dealt with privately, if at all; but today not only have these been taken over by the state, but we expect it to do so and seek to cast on its shoulders responsibility for still more. This paper also utters a reminder that the more we expect the state to assume obligations towards us, the more should we be careful not to lose our personal sense of obligation.

In the word "obligation" we have a juristic implication of prime importance at the very threshold of the inquiry. So let us begin by considering what it means. We owe to Professor H.L.A. Hart of Oxford the elucidation of part at least of this concept[3]. He distinguishes between "being obliged" and "having an obligation" and illustrates it with an example. If a gunman claps a pistol at X's head and demands his purse, X may "feel obliged" to hand it over, but he has no "obligation" to do so. "Having an obligation", says Hart, is the result of a rule to that effect; and, clearly, there is neither rule nor obligation that X should surrender his purse. I seek to go further and distinguish between "having any obligation" and "having a sence of obligation", which for the purposes of this paper is more crucial than the distinction between "being obliged" and "having an obligation".

If "obligation" depends on a "rule", we need to turn next to "rule", for the clarification of which we again owe much to Professor Hart[4]. He points out that "rule" connotes acceptance of a behaviour pattern as a standard for oneself and others. Failure to conform to it is condemned as "wrong" and conformity is "right". Such acceptance signifies the adoption of what he calls the "internal point of view" with regard to the behaviour in question, which he contrasts with the "external point of view" of an outside observer, who merely expresses what he sees descriptively, namely that "it is a fact that people behave in such and such a way". Internalisation results in more than giving descriptions. The point is not that people do or do not in fact conform to the behaviour pattern, but that they ought to do so, that is, behaviour has been accepted as a standard. With internalisation we move from the realm of the Sein into that of the Sollen.

Many rules prescribe patterns as to how people ought to behave, but not all of them are "lawyers law"[5]. Rules are identifiable as "legal" with reference to the criterion, or criteria, of validity in any legal system, that medium or media from which rules derive the quality of "law"[6]. In some countries this is a constitution, but in Britain, which has no constitution, the criteria of validity are Acts of Parliament, judicial precedents and (playing almost a negligible role) immemorial customs[7]. A rule emanating from a law-making medium is internalised as "legal" because the medium itself has been internalised, a point to which we shall come presently.

Internalisation, however, gives only a partial understanding of "rule" and "obligation". We cannot just stop there, but must go on to the next question: how and why do people internalise? The answer reaches into morality, history, sociology and psychology, which are severally interwoven into the "rule-ness" even of rules of law. Moral rightness is an obvious incentive to the acceptance of a law as a standard of conduct. Even in days gone by when people used to accept the utterance of some charismatic lawgiver, they did so because of their trust in his moral judgment, which was believed to have been divinely inspired[8]. At the other end of the scale there is internalisation through reason and reflection, which occurs when a person weighs up pros and cons and decides that he should accept for the general good, even if it goes against his own interest. Sociologists drew attention long ago to imitation as a basis for internalisation, that is, the herd instinct to do what most people do[9]. This is not unconnected with the first reason, since the fact that the majority behave in a certain way gives rise to the unthinking feeling that this must be because it is right.

Internalisation also occures because some patterns of conduct are means of achieving certain ends. Society consists of a network of interlocking practices of various groups, which becomes increasingly complex as time goes on. So much so is this the case that if one group refuses to play its role, other groups are stricken into impotence or at least inconvenienced. One has only to reflect on the extent to which the daily work and existence of every individual depends on others providing water, fuel, food, clothing, housing, transport etc. The result of such inter-dependence is that considerable pressure builds up on people to conform to their roles so as not to unsettle others, which thereby leads to acceptance of conformity to roles as "right" and non-conformity as "wrong". A good many interlocking practices do eventually crystallise into law. A cruder manifestation of the means-to-an-end connection is the acceptance of a rule of law as a means of avoiding unpleasant consequences. A distinction should here be drawn between internalisation of a rule as such and the decision by an individual in a particular instance whether or not to obey a rule, which he has internalised. With regard to disobedience in a given situation the fear of punishment can exercise a limited influence, but this should not be exaggerated. For when disobedience is unintended, for example, through carelessness, the threat of punishment is inoperative. Only when a person intends to flout a rule does fear of punishment enter into his calculation and, even then, it is balanced against the chance of escape, the severity of the punishment, the possibility of the law not being enforced etc. When, and only when, the law has been administered firmly and successfully over a long period does a psychological reaction to fear of punishment lead to eventual internalisation of a rule hitherto opposed. Professor Olivecrona has pointed out that fear is uncongenial to the human make-up, which endeavours to rid itself of it. If fear holds people back from doing or not doing what they desire to do or not do, there will come a time when a psychological reaction takes place to get rid of this ever-present fear by an inward acceptance of conformity[10]. In this way law could have an educative function in moulding values[11]. The likelyhood of success, however, is limited by the fact that when such an attempt to re-educate people goes against deeply-felt sentiments, for example, those of religion, race and so on, a psychological barrier tends to be set up in the minds of people against the conditioning influence of law. The lack of headway made by the desegregation laws in the southern states of America illustrates the point.

Lastly, and not least, the internalisation of a rule in many cases is induced by the psychological pressure exerted by the mere fact that it possesses law-quality. To say that such and such a thing is "law" is to strike an attitude towards it and to evoke one. This derives from the prior internalisation of the institutions and procedures of law making. As soon as we are told that this or that is in an Act of Parliament (or in Britain in a judicial precedent or immemorial custom), we accept it as a rule of law without even knowing of its content. The inquiry thus gets pushed a stage further back into the reasons behind the internalisation of the law-making institution, namely, the criterion of validity. If the Queen and all members of Parliament assemble at a royal garden-party and assent to a measure, it will not have the quality of "law" nor psychological pressure; but as soon as they assent to the same measure after its due introduction in Parliament and observance of the proper procedures, it will have law-quality. The reasons why a particular criterion of validity came to be accepted in any country are to be found in the morality and the historical and social conditions prevailing at that time. In Britain the Crown-in-Parliament, that is, the legality of Acts of Parliament, was accepted as supreme in 1689 after the revolution of 1688-1689, which accepted it in place of the supremacy of the monarch's legislative power based on the Royal Prerogative. A moral reason for accepting this change was that the Crown-in-Parliament was expected to provide a guarantee against the immoral abuses of prerogative power that had preceded it. English judges accepted the new institution despite the illegality of the change[12], partly because of general weariness after nearly a century of civil war, and partly through their own political shrewdness in not challenging Parliament at a time when it was in no mood to be challenged.

The effectiveness of a legislative body is undoubtedly a factor inducing acceptance of it as valid, but this may take some time. It needs to be stressed that effectiveness is <u>not</u> itself the basis of validity; it is only a powerful motive behind internalisation. The point can be seen in the Rhodesian revolution when the rebel regime of Mr. Ian Smith established itself effectively in power in 1965, but the Rhodesian judges unanimously refused to accept its validity until 1968; and British judges not even then[13]. Mere effectiveness did not make the regime legal; continued effectiveness was a factor which in time induced the Rhodesian judges to accept it. Usually, after a revolution the establishment of an effective authority and its acceptance as valid occur simultaneously, but this does not obscure the fact that effectiveness is only a reason for acceptance.

The other major criterion of validity in Britain, judicial precedent, was accepted not only because of the moral dictate of justice that like cases should be decided alike, but also because, in the absence of a code, the principle in a decision was not an illustration of some rule stated elsewhere, but was itself the only statement of law that existed. The validity of immemorial customs, which are always local variations of the general law, was accepted in the long past because of the need to do justice locally by fulfilling settled expectations and thereby fostering confidence is the, as yet new, centralised administration of law.

The foregoing should show it is not enough to say that "having an obligation" stems from internalisation of a rule and to stop there. For when internalisation is examined it is seen to result from a congeries of moral, historical, psychological, political, social and practical factors, so much so that it cannot be isolated from the context of law in society. So "having an obligation" and internalising a rule are more than matters of abstract law.

A further distinction was mentioned earlier between "having an obligation" and "having a sense of obligation". The latter is always moral and, as such, brings in considerations outside law. The two ideas are not congruent although they do overlap. If a legal duty is imposed on a person, he may internalise it to the extent of acknowledging that he is under a legal obligation, but if he has no sense of obligation with regard to the conduct in question, he may well be inclined to disobey. For example, every motorist acknowledges that to abide by speed limits is a legal obligation, but vast numbers of them have no sense of obligation with regards to it, which is why the law is disobeyed during every minute of every day. The imposition of a legal duty and the creation of a legal obligation are not conclusive; there always remains the underlying moral choice whether to obey the legal duty or not. A different illustration comes from cases where the law leaves people with liberties to act or not as they please, that is, without constraint either way. Freedom of action can be exercised abusively or with responsability. Since the law does not constrain, restraint on liberty cannot come from law. It can only derive from a moral sense of obligation. This is why the distinction between "having an obligation" and "having a sense of obligation" should not be overlooked. Law without a sense of obligation is unworkable. so any jurisprudential speculation about the nature and function of law must take account of this moral dimension.

Not only is the juristic significance of "having an obligation" and "rule" implicit in the imposition of obligations on the state in a Caring Society, but a sense of obligation is important to the success of such a society. There is a paradox in that the more we develop into a Caring Society, the less sense of personal obligation there seems to be in those spheres in which we expect the state to take over. In short, a Caring Society could produce uncaring people.

The explanation lies in a common feature of human nature. The Caring Society seeks to satisfy basic needs which few, if any, would have expected fifty years ago. Now, no one likes to be under an obligation of indebtedness or gratitude to anyone else. With some people, at any rate, the reaction to welfare benefits is to persuade themselves that there is no call for indebtedness or gratitude because all such benefits are due "as of right". Further, the satisfaction of "basic needs" implies that beyond them are other amenities, classed as luxuries, which are available to some but not others.

Thus there will be "haves" and "have-nots" in every society. There is a tendency for some of the have-nots, once they persuade themselves that the satisfaction of "basic needs" is only due to them as of right, to continue demanding the things which are not provided on the ground that these, too, are "basic" and hence due as of right. The more that is given, the more tends to be demanded. As Professor Hônoré observed, "What is at one time a luxury becomes at another time a necessity and need"[14]. Thus it is that a sense of obligation diminishes in inverse proportion to increasing insistence on "rights". A sad by-product of a noble endeavour.

Thus far we have considered "obligation" and "rule" as the first two juristic implications of the concept of a Caring Society. In "rights" we come to the third, and one which is no less important. "Right" is a word which is so emotively loaded that it has done more mischief perhaps than any other. Before anyone, layman of lawyer, can think and talk meaningfully about "rights", it is essential that its different connotations are distinguished. At this point an aspect of what passes under "analytical jurisprudence" can render valuable assistance.

"Right" is a homonym, which has undergone four shifts in meaning[15]. First, there is "right" in the sense of "claim", or demand, in one person that someone else should perform a duty. It is this correlative duty that imparts content to the claim. To say, for example, "I have a claim", is vacuous; but to say, "I have a claim that X should pay me his debt" immediately becomes meaningful. Distinct from this is "right" in the sense of "liberty", or freedom, to perform or not perform some act, that is, absence of a duty to act or abstain. Claim is passive; it is what the claimant demands that another shall do or not do. Liberty is active; it is doing or not doing something oneself. Correlative to liberty is an absence of claim, or "no-claim", in another person, which implies, of course, the absence of duty in oneself since liberty is absence of constraint. Thus, my "right" to wear a bowler hat is a liberty to do so; and there is my equal liberty not to do so. Liberty and claim often go together. For example, X's liberty to wear his bowler hat is normally accompanied by his claim against Y that Y should not interfere with him, correlative to Y's duty not to do so. The two concepts are, however, distinct, as when X gives Y permission to prevent him from wearing the hat if he can. X still retains his liberty to wear the hat should he succeed in eluding Y's attempt to prevent him; but if Y succeeds, X cannot complain since he has extinguished his claim while preserving his own liberty. The third connotation of "right" is "power" in the sense of legal capacity to alter the existing legal position of another or others, correlative to a liability in the latter to have his or their legal position changed. Powers pervade the legal system at all levels, from the powers of Parliament to make and unmake laws affecting everyone, to the power of an individual, for example, to accept an offer and so alter the legal positions of himself and the offeror by creating a new contractual claim-duty relationship. Fourthly, "right" means "immunity" in the sense of being in a position where one's legal condition cannot be altered by the exercise of a power by someone else, and correlatively the latter has a disability, that is, absence of power. The "right" of diplomats is an example: a diplomat cannot be affected by a power of legal action. Appreciation of these distinctions is essential to precise thinking. The so-called "right to make a will", for example, can be analysed as follows. The will itself is a power in that it alters the legal positions of various persons, executors, beneficiaries, etc.; there is a liberty to make a will, that is, to exercise this power, as well as a liberty not to make one, since one is free either way; there are claims against others not to interfere; and there is an immunity against being deprived of will-making capacity. Whenever the so-called "right to make a will" is under consideration, it is important to know which of these jural relations is in issue.

Applying this analysis to current slogans in social and political rhetoric, whether in a Caring Society or any other, it can be seen that declamations about "fundamental rights". "rights of Man", "Women's rights", "prisoners' rights", "animal rights", "workers' rights" etc. are ridiculous bases for serious action. Are we talking of claims, liberties of action, powers, or immunities; in whom; in what circumstances; when; where? The point can be illustrated by taking one of these, "workers' rights". Industrial action in support of these presents a sorry enough picture. When analysed, it can be seen that one aspect is the claim-right to be given jobs, correlative to a duty in the state to provide them, which deserves sympathy especially in a climate of unemployment. Next there is the assertion of an unrestricted liberty-right not to work at the jobs, which are being claimed "as of Right", that is, the liberty to strike. Side by side with this is the assertion of an unrestricted liberty-right to prevent other workers from working at their jobs by picketing. Going further there is the assertion of power to drive workers out of their jobs through the closed-shop doctrine. Finally, there is the assertion of immunity from suits in courts of law and immunity from having these liberties and powers curtailed or abolished. Reduced in this stark way, "workers' rights" present a bizarre picture indeed[16].

If these "rights" were to be realised in full, they could strike at the vitals of the society that is benevolent enough to indulge them, which would be a Caring Society, since they would receive short shrift in an uncaring one. The only way of avoiding the absurdities and anarchy that could result is by exercising restraint and responsibility in their exercise and assertion, which ties up with what was said earlier about the importance of a sense of obligation in a Caring Society. If it is not to prove harmful by making people morally flabby through an overweening insistence on "rights", they need to develop an increased sense of obligation so as to be worthy of the goal. Otherwise its achievement may not enure to the ultimate good of the people for whom it is designed. Is this an unpalatable insight, or just a truism?

One demand "as of right" made of any society, especially of a Caring one, is for better and better justice. Although traditionally it is thought of as a philisofical question, it has obvious legal overtones. For what legal system is there which does not aspire to justice as one of its most important aims? This, accordingly, is the next juristic, or quasi-juristic, implication of a Caring Society. The classic treatment of justice was by Aristotle, who began by distinguishing between "distributive" and "corrective" justice[17]. The criterion for the former was said to be equality: equal distribution among equals with the corollary of unequal distribution among unequals. "Corrective" justice comes into play when distributive equality has been upset, for example, by wrongdoing, and seeks to restore the status quo as far as possible.

Aristotle's analysis is an admirable starting point for discussion, but it is open to objection. First, his basic distinction between "distributive" and "corrective" justice is not clear-cut. Thus, in America today there is sensitivity that persons of African and Red Indian descent have been underprivileged in many ways, including educational opportunities. Accordingly, certain universities adopted a policy of admitting such persons with lower educational qualifications than were required of white applicants[18]. Is this kind of "reverse discrimination" an attempt at achieving distributive justice by treating lower educational qualifications of those who had lacked good educational facilities as equivalent to slightly higher qualifications of those who did have good educational facilities, or is it a form of corrective justice to redress unequal treatment in the past? Secondly, the justice of the formula, "equal distribution among equals", depends on the justice of the criterion by which equality is judged. If, for instance, sweets are distributed on the basis that fair-haired children shall receive two sweets each and dark-haired children one sweet each, perfect Aristotelian justice can be done. The question however remains: is it just to adopt that criterion for distributing sweets? All that his formula means is that according to the criterion adopted, one person is entitled to complain of injustice if he or she is treated less well than others if they are all classed as equals. This leads to the third difficulty: what if that person is treated better, for example, if one dark-haired child is given two sweets? Is this just or unjust as regards to that child? As regards to fair-haired children? The formula offers no answer. Fourthly, people are not in fact equal in many respects, intelligence, strenght, skills etc. For the law to treat people who are not in fact equal as being on an equal footing can in some cases create injustice. Traditionally the law of contract treated contracting parties as being on an equal bargaining position when often this was far from being the case. Employers, who in the past used to hold the bargaining advantage, were able to exploit workers by forcing them into unfair agreements, while even today the citizen is not on an equal footing with public enterprises and has no option but to agree to the terms of standard form contracts, or go without a necessary service. If the law were to take account of inequalities in fact, then it would have to mete out unequal treatment in law. The question when the law should depart from the criterion of equality cannot then be determined by equality. Finally, there is the question, equal distribution of what? Aristotle's answer was. "honours or money or the other things that fall to be divided among those who have a share in the constitution"[19]; which is hardly a workable basis upon which to organise any society, ancient or modern. Therefore, the conclusion from all this must be that justice is not synonymous with equality. Equality is only one aspect of it, an important one no doubt, but no more than that.

The mistake lies in regarding justice as some kind of "thing", which can be found and captured in a formula. On the contrary, it is a continuing series of tasks in fluctuating social situations. "Justice is never given", said Friedrich, "It is always a task to be achieved"[20]. I suggest five such tasks with the caveat that any list is a matter of personal choice. In my submission they are: (1) just allocation of burdens and benefits; (2) curbing abuse of power; (3) curbing abuse of liberty; (4) just decision of disputes; and (5) adaption to change. These tasks have been listed in this way for the purpose of exposition. They are not disjunctive for, as will appear, the second and third underlie the others.

Just allocation of benefits and burdens. Volumes can be and, indeed, have been written on this topic and different formulae have

been proffered. The vagueness of Aristotle's "honour or money or the other things that fall to be divided among those who have a share in the constitution" has been mentioned. Karl Marx said that in the proletarian dictatorship distribution will follow the maxim, "From each according to his capacity, to each according to his work", and in the communist utopia the maxim will be, "From each according to his capacity, to each according to his needs". In the first place, who decides which persons "have a share in the constitution"? For instance, there is a case in the Soviet Union in which a dissident mathematician is not being allowed to pursue his mathematics and has to shovel coal so as to avoid being classed as a social "parasite". Is his "capacity" to be gauged as a mathematician or as a coal-heaver? If it is usefulness to society that counts, how does one measure the value of a mathematician and a coalheaver, or a filmstar and a chef?
Again, who decides what are individual's "needs" and when they are deemed to be satisfied? The answer to these questions depend on the exercise of powers to decide them, which have to be vested in some person or body of persons. Underlying all this is the problem of curbing the misuse of such powers.

Assuming that a just scheme of distribution can be worked out, the next problem is that of winning acceptance of it by the populace and of keeping it acceptable. Society is not static, people's appetites change. Allusion has been made to the tendency that the more that is given, the more people may demand "as of right". Any scheme of just distribution will founder unless the liberty to campaign for more and yet more is restrained; which is why it was said that the twin problems of curbing the abuse of power and liberty underlie the others.

The phrase, "just allocations of benefits and burdens" is, in respect of vagueness, no improvement on Aristotle's, or anyone else's, formulation. Here the analytical distinctions in the meanings of "right" can be pressed into service so as to introduce a measure of precision. "Benefits" may be resolved into claims, liberties, powers and immunities; "burdens" into their correlative duties, no-claims, liabilities and disabilities. Such analysis, be it stressed, cannot of itself provide just distributions. It can only clarify the ground by revealing the problems, but ultimately the point will always be reached when a political decision has to be made.

Taking claims first, a distinction should be drawn between claims correlative to negative and positive duties respectively. The former abound in civil and criminal law, for example, the duty not to assault another, not to appropriate another's property, etc. With regard to these, the axiom should be that everyone should be equally entitled to such claims, unless a particular individual has voluntarily forfeited the claim, for example, by consent or aggression. Positive duties require a further distinction between claims against individuals and against the state. The former present no difficulty since they are voluntary created, for example, by contract where, in the absence of deceit, mistake and so on, a contracting party cannot complain of injustice merely because he or she has made a bad bargain. Positive duties in the state, that it should, for example, provide welfare benefits, must rest on state policy and in this connection what was said about demands "as of right" is relevant.

With regard to liberties, the problem of justice is different. It is not, ought these to be shared equally? but, ought some liberties to be allowed at all, such as a liberty to kill, maim or damage? and, how can the abuse of liberties which are allowed be curbed? All such matters have to be decided on policy grounds; equal distribution does not come in at all.

Powers present similar differences. Some of them, such as the power to legislate, power to decide what shall be the criteria of distribution, what are "needs" and when they have been satisfied and so on, obviously cannot be distributed amongst everyone. They have to be vested in some person, or body of persons, and the question of justice concerns the way in which the abuse of such powers can be prevented. On the other hand, in the case of private powers, justice requires that a good many should be distributed amongst all, subject to necessary restrictions such as the denial of testamentary power to the insane.

Immunities are in a different position in that justice requires the removal of special immunities, not equal conferment. In Britain the Crown used to enjoy certain special immunities; for example, it could not be sued in delict. The development of English law has progressively removed these. The example just given no longer obtains since in 1947 the Crown Proceedings Act made the Crown liable in delict[21]. Again, public authorities used to be protected against the power of legal action by specially short periods of limitation, but these too have now been removed. It would seem, therefore, that it is not equal distribution, but the conferment of special immunities that require justification. In the current industrial situation serious thought needs to be given to the justification of immunities from suit insisted on by trade unions. The question of justice here depends on one's stance in the political spectrum.

The distribution of burdens can be dealt with more briefly. Duties are created or recognised on grounds of policy, governmental or judicial. Justice here depends on a balance between a number of often conflicting considerations and no general formula is of any use. It is entering into cloudcuckooland to talk about imposing no-claims, which, as will be remembered, denote the positions of persons who have no claims that other people shall perform duties. The latter are therefore at liberty to act as they please. The question of justice thus focuses on the exercise of liberties, which will be discussed presently. Liability denotes the legal position of a person which can be altered by the exercise of a power by another. The existence of these, like duties, depends on policy. For instance, policy determines who shall be liable to income tax and on what scale, there is universal liability to rationing in times of crisis or shortage, and so on. Equality should be a very rough norm so that special variations in the imposition of liabilities require justification. Finally, in the case of disabilities, one aspect of justice requires their removal, not imposition. In the past large sections of the population suffered legal and political disabilities, such as women, Roman Catholics, Jews, dissenters, and even now racial groups suffer in countries with racial policies. Most of these have happily disappeared, but there is a long way to go yet. The imposition of special disabilities cannot be based on equality, but on special policy grounds. Should a person be allowed to insure privately against his own wrongdoing? Whenever this is wilful, it is considered contrary to public policy, stemming from moral considerations, to allow him to so so[22]. This has nothing to do with equality. When a person's wrongdoing is only careless, he is allowed to insure, since any moral consideration to the contrary is outweighed by the need to ensure that the victim receives compensation.

Curbing abuse of power. Power can mean physical force or the juristic ability to change the existing legal position. In either sense power, however great, is inert in itself; it is the exercise of it that has effect for good or ill. Such exercise is a matter of liberty. Power in the legal sence is the concern of the present discussion.

The most widespread and influential powers are those vested in government, and no one needs to be reminded of countless examples in ancient and modern history of their beneficial as well as detrimental exercise. A Caring Society can be consistent with unlimited governmental power as long as this is exercised with compassion and for good. Unfortunately, the tragedy of history has been that power seems almost fated to end in its abuse. As Lord Acton, the English historian, observed, "power tends to corrupt, and absolute power corrupts absolutely"[23]. The progress of human societies has spiralled between revolt for freedom from the power of an oppressor and abuse of the freedom so won, which eventually calls for a return to power so as to end the resulting anarchy. The ambiguity in the word "freedom" has been disastrous. It connotes "freedom from" power (in the sense of immunity) and "freedom to" act as some pleases (in the sense of liberty). Analytically, the distinction is obvious and had it been ingrained in common speech and thought a good deal of unhappiness might have been avoided. Failure to appreciate it has led champions of freedom from abusive power into the non sequitur that this freedom (immunity) also implies freedom to act without constraint (liberty). Abuse of liberties of action can then reach such a pitch that people, weary of the instability that results, call for a return to power to restore order at the expense of the liberties that have been abused[24].

It is not that power-holders or those striving for freedom from power are necessarily evil. Despite most laudable intentions, they get caught up in toils too strong for them. Thus, revolutionaries against an oppresive power-structure begin by seeking genuinly to establish a milder regime with wider liberties for those who had been oppressed. Their opponents, and there will always be some, then utelise to the utmost the liberties allowed to them to agitate against the new regime. Understandably, those who have just won power cannot allow the fruits of their achievement to be destroyed in this way, so they find themselves forced to exercise power to restrain and even outlaw the agitators. Having succeeded in forcing the previous power-structure to yield to them, they know well at what points agitation can succeed and will thus be all the more effective in blocking those loopholes for liberty of action against themselves[25]. This is how those who rebel against an oppresive regime often end up by establishing a regime al least as oppresive, if not more so. The Soviet revolution is an example. The longer they are in power, the more resistance they will encounter and tighter will become the restrictions. Power does indeed corrupt, and absolute power corrupts absolutely. If a Caring Society were to be in the hands of the "philosopher-kings" dreamed of by the Greek thinkers, there would be no danger of abuse of power by such enlightened beings. Actual human beings are different. Power can only get into the hands of determined persons as a rule. They are often unscrupulous, but even if they are honourable, they are, to some extent at least, in the grip of less honourable persons who helped them to their positions.

Some exercise of power there will always have to be. Hence it becomes all the more important for Caring Societies to evolve ways and means of curbing abuses of governmental power. Broadly speaking, abuse of such power manifests itself in unjust laws. The problem is how to deny validity, that is, the quality of "law", to a properly passed but unjust enactment. St. Augustine denied that it was "law": <u>Lex esse non videbitur quae justa non fuerit</u>[26]. More recently a judge of the House of Lords in Britain said, "To my mind a law of this sort constitutes so grave an infringement of human rights that the courts of this country ought to refuse to recognise it as a law at all"[27]. The validity of an enactment normally rests on a purely formal requirement. So, if an Act of Parliament has been duly passed, it is commonly treated as "law" no matter what its content may be. To invalidate an unjust enactment that has been formally passed, its validity will require two requirements, the formal one plus a minimum quality of justness. A Pakistani judge expressed this when he said, "A law is not law merely because it bears that label. It becomes law only if it satisfies the basic norms of the legal system of the country (enshrined in the Qu'ran) and receives the stamp of validity from the law courts"[28].

Many reasons can be adduced in support of the need for such control. First, pure formalism enables oppresive regimes to arise and consolidate themselves by passing oppresive enactments which satisfy all formal requirements. The establishment of the Third Reich is an example. Secondly, formal criteria are themselves based on the assumption that what is enacted will be just. If not, formalism itself becomes suspect. "The capricious orders of a crazy despot", said Sir Frederick Pollock, "may be laws according to Austin's definition until they are revoked; but if so, it is worse for the definition"[29] Thirdly, rule by formally enacted laws is a betrayal of the individual if all it means is the impartial adminstration of formally valid, but oppressive and degrading laws. Fourthly, formal criteria were themselves adopted originally because they were thought to guarantee some quality of justice, as in Britain where the formality of the Crown-in-Parliament was adopted in place of the Royal prerogative in order to guard against abuses of power as experienced under the prerogative. Fifthly, the just quality of laws enhances fidelity to them. Sixthly, it is not correct to say that nothing is gained by denying the label "law" to an unjust enactment. "Law" has powerful emotive connotations and the refusal to attach it to a particular enactment can have significant psychological effect. Finally, laws are designed to endure over long periods. Just quality is a condition of continuity since without it an enactment is doomed to ultimate futility, whether by way of repeal, disobedience or disregard.

The practical problem of how to introduce a minimum just quality into the requirements of validity encounters formidable difficulties. First, whose moral sense should prevail? It cannot be that of legislators, since the object of the exercise is to protect society from the immoral values of legislators. Nor is it that of society at large, which is too diffuse to be consulted readily and, in any case, is likely to be evenly divided to be controversial issues. It had to be some objective test, but the chances of securing general agreements on such a one are small indeed. There are many Declarations of Human Rights, Inalienable Rights, Fundamental Freedoms, littered through history, any one of which could serve as a model. The following considerations are offered as guidelines in framing such a charter. (1) Power should not be exercised exclusively for the benefit of the powerholders. (2) Power should not be exercised so as to exclude, or accord unequal treatment to another for reasons of race, religion or opinion without special justification. Such justification should not infringe any of these guidelines. (3) Power should be exercised according to accepted procedure, that is, after due process[30]. (4) Holders of power should not be above having to conform to their own dictates. (5) Holders of power should not become a closed circle. This concerns participation in power, and in deciding who should be allowed to participate the following further considerations are relevant: (a) the permanence or otherwise of the involvement, as in a university or in the state; (b) voluntariness of the involvement so that, where it is voluntary, conditions could be imposed; (c) the relationship between the society and its members; and (d) the abilities of would-be participants to contribute.

Assuming now that a minimum test of just quality can be envolved, the next difficulty is crucial. How can it be made practicable? Consider first how it could be given effect through the judiciary. At best this is of limited scope and has many drawbacks. There are four ways in which it might operate. First, when interpreting statutes judges do utilise the ambiguities of words so as to minimise the effect of harsh legislation. Secondly, they could invoke certain principles to invalidate certain abusive exercises of executive, as distinct from legislative, power, for instance that a particular action was not in accordance with the principles of natural justice[31]. Thirdly, more rarely, a court may treat the guarantee of justice on which a particular legislative authority had been accepted initially as a built-in limitation on its power. For example, Article 1 of the Bonn Basic Law declares that the German people acknowledge fundamental human rights, and the succeeding Article provides that the legislature is powerless to contravene them. The Federal German Constitutional Court declared that "They are referred to in the Grundgesetz as inviolable and inlienable human rights; not as guaranteed by the Constitution, but as existing before it and independently of it"[32]. A fearless Pakistani judge invalidated a decree of the government on the ground that that government had been accepted initially for the purpose of "restoring sanity" and law and order and that "No one, including the Chief Martial Law Administrator, can transcend or deviate from the sole purpose of restoring law and order and democracy and it needs no gainsaying that curbing the jurisdiction of the established judiciary is not a step in that direction"[33]. Fourthly, judges might be able to invalidate a statute on the ground that it had not conformed with the accepted manner and form for legislating.

Against all this, the objections to judicial control are formidable. It requires an independent judiciary brave enough to challenge an all-powerful government, which is likely in any case to get rid of all persons standing in its way. The cynical English philosopher, Thomas Hobbes, put it bluntly: "In the matter of government when nothing else is turn'd up, Clubs are trump"[34]. Next, on what test will the judges act? A controlling set of just principles can only be evolved through case-law, if at all, over a long period, which is dilatory, and the accidents of litigation are too haphazard and irregular to be of service. Finally, the decision to declare an injust enactment void will be a political decision, and the judicial processes of fact-finding and rule-applying are not for this purpose.

Another way of checking the power of legislatures is by inserting a disability into the basis of legislative competence. The possibility of judges recognising such an in-built historical limitation has been mentioned. Hitherto the most effective way of imposing disabilities has come via written constitutions. Where these exist, courts can strike down legislation as being unconstitutional. Much depends, however, on the precision or vagueness of the wording of the constitution. Usually this is vague enough to enable a sympathetic or unsympathetic judiciary to uphold or avoid statute[35]. Also, the ease with which a constitution can be amended will be a measure of the protection which it can provide.

Finally, the machinery of the European Common Market points the way towards a supra-national means of controlling the power of national legislatures. Within its sphere of operation, any national legislation contrary to the Rome Treaty is void to the extent of conflict. If it were possible to establish a Human Rights Treaty with machinery comparable with that of the Roman Treaty, it would become possible to create a check on the power of national legislatures to enact unjust laws and no country will be able to alter or amend that limitation. This is still a dream of the future, but at least the Roman treaty had made a reality, albeit in a modest way, of something that would have been inconceivable fifty years ago.

Curbing abuse of liberty. Abuse of liberty is not the road to liberty. People have struggled for liberty throughout the ages since "liberty is ancient; it is despotism that is new". The problem of curbing the abuse of power showed the limitations of law in this regard. This, however, is minor compared with the problem of curbing the abuse of liberty since it cuts deeper. The point has been made that power as such is inert and that it is the exercise of it in harmful ways that needs to be curbed. Such exercise consists of the twin liberties to do so or not, which underlie power. With liberty law reaches its limit, for even though a legal duty may be imposed with regard to some action, there still lies beneath it the inner moral liberty to obey or disobey it.

Dealing first with the prohibition of liberty, this is achieved through the imposition of duties forbidden the doing of certain acts, or enjoining the doing of other acts. It is impossible to impose duties on sovereign legislatures not to exercise power in any way they choose because they are sovereign, unless this could be achieved through some futuristic supranational authority the possibility of which has been discussed. Sovereign legislatures can of course impose duties on their subject so as to curtail the latter's liberties. Such prohibitions result from political decisions the justice of which depends on values of various kinds. The vast majority of societies forbid murder, robbery and a variety of acts deemed criminal or delictual. Liberties to act in this way simply cannot be allowed. There are other more arguable kinds of action. For instance should the liberty to exercise economic power by way of monopolies, profiteering, or exploitation be allowed or forbidden? Some countries do prescribe certain forms of these. Should the liberty to use industrial muscle by way of strikes, lock-outs, or picketing be allowed or forbidden?

No less controversial is the relationship between law and moral liberty. How far should laws uphold positions? First, the significance of "should" needs explanation. Some laws reflect moral positions, while others do not. The question is not whether laws do or do not reflect morality, but whether they should do so, that is, the moral justification for using laws in this way. One answer is that moral tolerance, which is perhaps associated especially with Caring Societies, could reach a point when society fragments into a number of so-called "alternative" societies with their own codes and practices. Tolerance implies either the existence of a moral majority that tolerates minorities, or mutual tolerance and co-existence between a number of groups. The latter loosens cohesive society; which is even so pronounced a positivist as Professor Hart felt driven to admit that "some shared morality is essential to the existence of any society"[36]. So in order to preserve the very existence of society, some reinforcement of morality by law is essential.

The main point at issue is the interpretation of "moral position", and this lies at the heart of the debate between Lord Devlin and Professor Hart. According to the former, every society evolves moral institutions, which become part of its fabric[37]. Thus, monogamy is fundamental to Christian societies, but not to others. Therefore, he argues, society is entitled to protect its fundamental morality, and thus itself, against anything capable of destroying it. He is not against change in moral ideas, only that change should be gradual. Professor Hart, on the other hand, contends that laws should not uphold morality as such, but should only prevent harm to others as this is destructive of any society[38]. It would seem that they are not arguing about the same thing. Lord Devlin is referring to the morality which is part and parcel of this or that particular society, whereas Professor Hart is referring to the morality which is essential to the existence of any and every society. Both are open to criticism. Lord Devlin's test for the morality which a society is entitled to protect by law is the reaction of the ordinary man; which is difficult to discover and too vague to provide a viable basis. On the other hand, Professor Hart's position is actually based on a moral premise, namely that it is morally wrong to use law to uphold morality as such. "Harm to others" is necessarily wider than physical injury, and he justifies such wider use of law on ground of "paternalism", that is, the concern of the state to prevent people from harming themselves, an assertion which goes some way towards undermining his own position. For instance, he supports the use of law to forbid homosexuality with youths because of the paternalistic concern of the state to protect them from what he calls "corruption". What is "corruption" but a moral judgment, namely, concern for their moral welfare? Here, then, he is supporting the use of law to uphold a moral position as such after all. He also supports the legal prohibition of bigamy as being a form of public nuisance and an affront to a public ceremony and because it throws legal relationships into confusion. In this example he is making Lord Devlin's point in that affront and confusion stem from the fact that monogamy is part of the fabric of _this_ kind of society.

Not only are they talking about different things, but Lord Devlin fails to provide a workable test as to when legal reinforcement of morals should begin, while Professor Hart fails to expel morality at the justification for some laws at least. The answer to the problem does not lie in formulae such as "harm to others", or any other. It is necessary to consider further the machinery of the law that might be used, and this is where the methods by which moral positions might be "upheld" come in. Much can be accomplished by attaching disabilities rather than imposing prohibitive duties. Thus, even though the prohibition of adult homosexuality is removed, as it has been in Britain, the moral disapproval of it in law can still be envinced by attaching disabilities, for example, to make valid contracts for such purposes, or to sue *ex turpi causa*[39]. Also, other collateral duties can be preserved. Thus, it is still the offence of conspiracy to combine in order to urge homosexual practices[40]. Again, in the law of evidence an allegation of homosexuality is an imputation against good character[41]. When considering direct prohibition through duties, several factors should be taken into account among which are the danger of the activity in question to others, economy of forces for detection, possible hardship of the penalty, and also the appropriateness of the civil or criminal law, or perhaps of some special machinery. In the light of all this it is clear that there is no simple solution.

So far the discussion has concerned the use of law to curtail liberty. Where the law allows people liberties of action, restraint on their abuse obviously cannot come from law; it can only result from self-restraint and self-discipline. Law had reached its limit and the only question is whether it can lend indirect assistance towards promoting self-restraint and self-discipline.

Liberty at law may safely be allowed where there is assurance of a sufficient measure of moral discipline. What has to be done, therefore, is to preserve that "shared morality" referred to by Professor Hart. Hitherto this has come via religion, but religion today is fast losing its hold. To relax legal restraint at a time when moral discipline is on the ebb is the path to eventual social desintegration. One way of preserving the "shared morality" is by restoring the authority of religion in the broadest sense, not just of Christianity. How far the law and its techniques of statute interpretation and case-law, rules of evidence etc. might help towards a revised and acceptable understanding of religious texts is a fascinating topic which lies outside this paper.

A shared morality implies a moral atmosphere and certain accepted standards. In so far as laws do embody standards, an indirect way of preserving the shared morality is by not repealing laws embodying moral standards merely because they are difficult to enforce. To abolish such a law for this reason is to relax at one stroke the pressure to obey other enforceable laws. Obedience does rest, however, slightly, on the psychological pressure exerted by something being "law" and also on the fact that some people continue to obey. Repeal because of disobecience or difficulty of enforcement not only demoralises the law-abiding, but is also a step towards undermining hitherto enforceable laws and so leading to their repeal as well by encouraging widespread disobecience to them.

Attention was drawn to the point that even where there is a law imposing a duty, an individual has an inner moral liberty to obey or disobey it. Restraining the liberty to disobey is perhaps the most important problem of all. If a Caring Society aims to be libertarian, it needs to work out the limits within which disobedience is to be tolerated. A popular thesis today is that laws rest on consent, which cannot be accepted at face value. No one is asked in advance whether he or she agrees to be bound by law, which applies automatically. The question is not whether law is binding because of consent, but whether subsequent withdrawal of consent makes a law cease to bind. In this event the organised force of the state can be called into play, which makes consent or its absence irrelevant. The question then is when it is justifiable to challenge the organised force of the state and when it is justifiable for the state to use force. The following limits for tolerable disobedience are suggested. (a) Obedience should be the norm so that disobedience needs to be justified. (b) Disobedience may be justified to save life and, arguably, property. (c) Available means of redress must be tried before resorting to disobedience, and in this connection the likelihood of success of such means is relevant. (d) Disobeciende should not impair the equal liberty of others to obey. (e) Disobedience to provide a test case is acceptable. (f) Disobedience as a plea for consideration or reconsideration of some decision should cease once consideration has been given, even if the result is to confirm the original decision. (g) Disobedience is persuasive when designed to gain publicity and a hearing when other means have failed. The danger of this is that the public gets tired of such demonstrations so that, in order to hold public attention, they intensify. (h) Disobedience should not involve hardship on others or violence. "Hardship" includes the breakdown of social existence, social services, or undue expenditure in containing or repairing the effects of disobedience. (i) Disobedience of a national law in order to protest against some evil in a foreign country, especially when it inflicts hardship on uninvolved nationals, is not permissible.

Increasing disobedience and abuse of liberty reflect a diminishing sense of responsibility. It was pointed out earlier that a possible by-product of Caring Societies is the promotion of a sense of "rights" rather than of obligations and how important it is to shift emphasis to the latter. Another disservice results from the well-meant efforts of psychologists, who take pains to explain that disobedience and delinquency are attributable, not to the delinquent's shortcoming, but to the shortcomings of others, the "environment", and so on. Responsibility surely lies on both. To foist responsibility away from the delinquent and on to others in the past is logically to accept <u>now</u> the delinquent's responsibility for the future. Responsibility cannot be evaded. The concern of psychologists is too onesided. Compassion: by all means, yes; but at the expense of standards and responsibility, no.

Champions of liberty, be it noted, adopt essentially a negative approach, namely, doing away with restraints. A positive approach would pay heed to what people <u>do</u> with their liberties and would start with people's duties to society and work outwards from these to liberties. Law, it should always be remembered, serves society.

<u>Just decision of disputes.</u> Every kind of society aims at the just resolution of disputes, and it is the task which lies closest to the work of lawyers. No judge, assuming office, swears simply to apply the law because this is not possible. As Mr. Justice Holmes, the great American judge, put it: "General propositions do not decide concrete cases"[42]. The structure of rules is such that they allow an element of discretion, large or small, in their application. In similar vein Lord MacMillan once said, "In almost every case, except the very plainest, it would be possible to decide the issue either way with reasonable legal justification"[43] and Lord Wright, too,

said, "notwithstanding all the apparatus of authority, the judge has nearly always some degree of choice"[44]. Nor does a judge swear simply to do justice as he feels. What he undertakes to do is justice according to law. Since rules do not of themselves decide disputes, the inspiration guiding the way in which they are applied must come extra-legally. It comes from values, which shape a judge's sense of justice and by which the competing interest in every dispute are measured. Lawyers are so familiar with the techniques and methods of reasoning involved in this process that any treatment of them here is unnecessary.

Adaption to change. This is the last of the five tasks of justice enumerated at the beginning. Society is not static: changes in the environment, ideas and outlook have ultimately to be reflected in law. Failure to adapt to change is in its way as much an abuse of power as direct exercises of it. No legal system should aspire to an ideal of unalterability like the Laws of the Medes and Persians, for it will end as they did -in extinction. The pressures behind change are manifold. There is the slow evolution of customary practices over long periods of time; the technological explosion has revolutionised work practices in all directions and computers especially are rapidly making many rules of law and evidence outdated; sex-change operations have thrown into confusion settled ideas as to what a "man" is, or a "woman"; life-support machines have made uncertain what "life" is and when "death" occurs; and so on. The techniques by which law can keep abreast of changes are principally legislation and judicial tinkering day by day with the scope and meaning of rules and concepts. The importance of it all can hardly be over-estimated, but the topic of law reform moves into areas outside this paper.

Concept of law for a Caring Society. It should be evident even from the little that has been said in this short paper that "legal philosophy" is only a part of what is understood as "jurisprudence" in Britain. Law is a social institution and jurisprudence concerns the study of law in society. It is thought about the nature, function and funtioning of law on the broadest possible basis, and about its adaption, improvement and reform. The wide sweep of the subject can be brought home by glancing at the implications of this description.

The word "thought" lays an essential foundation for study in that it deals with the meaning of words and language, since thought is shaped by language forms. "The lawyer's business is with words", said Lord MacMillan. "They are the raw material of his craft"[45]. Laws are articulated in words, logic, as the very name implies, is governed by words, different types of argumentation, rhetoric and much more besides, all depend on the use and manipulation of language. A sharpened awareness of the potentialities of language to lead and mislead thought is conducive to clear thinking to the disposal of many problems and controversies as purely verbal and a waste of energy. The overall aim of this branch of philosophy, known as semantics, is to teach people how to think rather than what to know.

The study of substantive law informes one of the various propositions of law that comprise various branches, contracts, property, etc. Jurisprudence consists of speculations and generalisations about law. Obviously, without at least some knowledge of what the law is, it is not possible to speculate about it. Granting such knowledge, there is no limit on the different kinds of speculation. The nature of law in one form or other is common to all "jurisprudes". Law is not of uniform texture and consists of varieties of legal material, and one kind of study devotes itself to examining these. Then there is the conceptual apparatus, such as rights and duties, ownership, possession and so on. The process of doing justice according to law in deciding disputes could be assisted by a study of the conceptual tools used in legal reasoning and by keeping them adapted so as to cope with changing social problems. There is also the institutional apparatus of the law, such as the institution of contract, property, as well as institutions like courts, police, and so forth. Much attention has also been paid to the nature of "a law" and "law" in the sense of "legal system". Some jurists have believed that it is necessary only to identify "a law" since a "legal system" is the sum total of laws[46]. This, it is submitted, is not so. A concept of "a law" presupposes a concept of "legal system" and, moreover, "a law" is not an objective phenomenon, but only represents the way in which a particular individual has chosen to classify certain material[47]. Be that as it may, most laws are prescriptive of behaviour, that is, they do not describe how people actually behave, but prescribe how they ought to behave. A law prescribes that X ought not to steal, but there is no contradiction in adding that X does in fact steal. The two statements operate in different spheres of discourse, that of the Sollen and the Sein respectively. The idea of "ought", which is implicit in prescriptions, reflects the aim or end sought to be achieved. This opens up the function of law in the sense of purpose. Laws and legal system fulfil many functions of which, as has been suggested, the achievement of justice is one of the most important. This has been touched on and no more will be said. The functioning of law deals, on the one hand, with legal processing in applying and enforcing the law, and, on the other, with the actual ways in which law operates in society. The latter study enlists the aid of sociologists with their inquiries into the interrelationship between law and social life.

It is clear, therefore, that jurisprudential study reveals how inextricable is the connection of law with other disciplines: it is the philosophy of law, the politics of law, the sociology of law, the morality and the ethics of law. The legal philosophies of great thinkers are only a part of all this, an important part no doubt, since much of this store of wisdom is helpful in solving the problems of today. In sum, jurisprudence seeks to depict law in life, not just law in books and, above all, it teaches that no one can be a good lawyer who only knows law.

All these aspects of jurisprudence are implicit in a concept of law for a Caring Society. In the first place, what is needed is not a concept of "a law", but of "law" in the sense of "legal system". A legal system is a co-ordinated and purposive activity. It is more than the sum total of individual laws. For one thing, it embodies the pattern of interrelation of its laws. A railway system, for instance, is not the sum total of the tracks and rolling stock laid side by side, but the pattern of their lay-out depending on the concentration of natural resources, location of ports, hills, plains and so on. So, too, a legal system has its own lay-out of laws. The co-ordinating factors are validity, which imparts the quality of "legal" to all its parts, the conceptual structure, which draws together groups of laws and categories of situations, and the institutional structure with which the law is administered.

A legal system also differs from individual laws in the scale of its purposive function and functioning. this introduces an important consideration. Every functioning activity must take time over its functioning; from which it follows that every functioning phenomenon, together with its component elements, must continue to exist over that period of time. Therefore, all conditions essential to continued existence and factors relating to functioning are implicitly part of the concept of it as a functioning phenomenon. There are thus two conceptual frameworks in which one may view any phenomenon, that of the present moment here and now and that of a continuum of time. They are not mutually exclusive for both are relevant in different situations. Perhaps, an analogy may help. We are prone to think of objects in terms of three dimensions, namely, length, breadth and depth. We regards a chair or a person on that way. The point, however, is that a three-dimensional concept is only that of an instantaneous chair or person, which has length, breadth and depth for a moment and vanishes. In the world of reality there are no such things as instantaneous chairs of persons; real chairs and persons exist over periods of time. This brings in the dimension of endurance. The continued existence of a human being depends on certain essential conditions, such as a minimum intake of calories, oxygen, parameters of temperature, of states of health and so on. Normally we are not conscious of these factors vital to living persons, but they are implicit all the time. When regarded in this way it is evident that our familiar three-dimensional concepts are only abstractions, that of instantaneous entities, which do not exist except in imagination. Concepts of reality have to be four-dimensional with all that that implies. For many purposes, no doubt, three-dimensional concepts, abstract though they are, suffice. For example, in order to accomodate persons round a table, the concept of each person as a three-dimensional entity is all that is necessary; the factors that make for his or her continued existence are all but superfluous. In a situation where the continued existence of persons is of moment, for example, in a hotel or a spacecraft, the factors essential to existence become vital.

One may similarly think of laws as things of moment for everyday purposes, the only relevant question being whether they bear the stamp of formal validity; or one may think of them as continuing and functioning phenomena, which then imports other considerations vital to continuity and bound up with functioning. The point of all this is that a legal system in its very nature can only be thought of in a continuum and much of what has been said earlier concerns the conditions essential to its peaceful continuity. Only in this way is it possible to combine formal validity with moral, ethical, political and sociological considerations within a single conceptual frame. Such would be a concept of law for a Caring Society, and it is within such a one that lawyers are able to address themselves as lawyers to the many important matters that are needed to make it caring.

1. The author would like to express his thanks to members of the audience, senior and junior, who loyally attended the lectures at Tilburg University in March, 1984, on which this paper is based. Some aspects of it were also the basis of an address to the Jessel Society in London in 1982.

2. If it is asked why a jurist should venture into sociology, the reason is similar to that which induces many sociologists to pronounce on law: law is a social institution.

3. The Concept of Law.

4. Op.cit., pp. 80 sqq.

5. E.g., rules of a game, wearing black at funerals, etc.

6. Called the Grundnorm of a legal system by Hans Kelsen.

7. After Britain's accession to the European Economic Community as from 1 January 1973, Regulations and Decisions of the E.E.C. have been given automatic law-quality in Britain. They operate only within a limited sphere and are not germane to the present discussion.

8. E.g., Moses and the Ten Commandments, The mistes in Homer, Anglo Saxon Dooms.

9. Tarde, Les Lois de l'Imitation.

10. Law as Fact (1939) pp. 147-153; (1971) pp. 271-273. See also Schwarz and Orleans, "On Legal Sanctions" (1966-67) 34 U.Ch.L.R. at p. 300: "The threat of sanction can deter people from violating the law, perhaps in important part by inducing a moralistic attitude towards compliance".

11. Cf. Aristotle, Ethics, II, 1.5: "Legislators make citizens good by forming their habits"; Lord Simon of Glaisdale in Charter v. Race Relations Board [1973] A.C. 868,,900: "I would only add it seems also to have been within Parliamentary contemplation that the law might perform in this field one of its traditional functions, an educative one - namely, to raise moral standards by stigmatising as henceforward socially unacceptable certain hitherto generally condoned conduct". The U.S.S.R. is trying to carry out the largest attempt to educate the masses in the values of communism by ruthless enforcement of law and intensive propaganda.

12. It was illegal because the legislative omnicompetence of the prerogative was legally unchallengeable; it was the basis on which the whole of the Common Law and system of courts had been built up. Parliament had only ever been consulted as a matter of choice. When test cases of prerogative legislation came before the court during the 17th century, the judges had to uphold its validity although they themselves were not always in sympathy with the King. Parliament's insistence that

legislation was invalid without its participation thus had no legal justification. The change was revolutionary because it was un-supportable on the basis of legality that had prevailed up to 1688. After 1689 a fresh start was made on a new basis.

13. Madzimbamuto v. Lardner-Burke 1968 (2) S.A. 284 (Rhodesian Appellate Division); 1 A.C. 645 (Privy Council in Britain).

14. "Social Justice" in Essay in Legal Philosophy (ed. Summers), at p. 78. The poet Oliver Goldsmith lamented the fate of a land "Where wealth accumulates and men decay": The Deserted Village.

15. W.N. Hohfeld, Fundamental Legal Conceptions (ed. Cook),Chap.1. The Hohfeldian distinctions and terminology are adhered to in this paper.

16. See my paper, "Götterdämmerung: Gods of the Law in Decline" (1981) 1 Legal Studies, at pp. 8-9.

17. Nichomachean Ethics, V.3.

18. In De Funis v. Odegaard, 40 L. Ed. (2d) 164 (1974), a disappointed white applicant sued the university for unjust discrimination, but the Supreme Court avoided the issue because he had been admitted before the final hearing. In the Bakke case, 438 U.S. 265 (1978), the Supreme Court by a bare majority held that the university was in the wrong to have discriminated, but finding a coherent principle from the judgments eludes ingenuity.

19. Op. cit., V.2

20. "Justice: the Just Political Act", 6 Nomos, Justice, at p. 34.

21. The reigning monarch, however, continues to be personally immune from liability by virtue of section 40(1). This emphasises the distinction between the legal personality of the Crown and of the monarch.

22. Geismar v. Sun Alliance and London Insurance Ltd. and Another [1978] Q.B. 383.

23. Historical Essays and Studies, Appendix. The same point was made earlier by Montesquieu, L'Esprit des Lois, 11, 4-6.

24. Thus, Greek democracy was established as a reaction from the power of the Draconian regime, but democracy was wrecked by the unruly Athenian mobs. This led Plato and Aristotle to preach against unlimited freedom and for the need to have an enlightened powerstructure. The chaos of the Dark Ages was succeeded by a powerstructure of a sort in the Holy Roman Empire. Successful agitation against this resulted in the rise of municipial states. Freedom from this power was taken to mean unlimited freedom to act as every state pleased in their mutual dealings, which led to the chaos of the Thirty Years

War. This prompted Grotius and others to call for a restraining power-structure in the form of international law. Abuse of domestic power by national sovereigns led to revolutions from it in England in 1688 and the French Revolution. The French revolutionaries so abused their freedoms of action that there was a return to power under Napoleon. Reaction after World War I set the permissive Weimar Republic in Germany to the brink of anarchy from which she was pulled back by the return to power under Hitler. The Russian revolution from Tsarist power has resulted in the most ruthless power-structure yet seen and the suppression of freedom, not only in the Soviet Union, but also in the satellite countries, e.g., Hungary, Czechoslovakia and now Poland.

25. E.g., in Pakistan the chaos into which political factions plunged the country after allegedly rigged elections was ended on 5 july, 1977, by a military regime, which promised free elections on 16 october of that year and a return to democratic government. Opponents of the regime utilised their freedoms to foment such unrest that on 1 october it was announced that elections were to be postponed indefinitely. Since then the government, which meant well at first, has found itself having to impose sterner and sterner measures until it is in the same unenviable position as its predecessor.

26. De Libero Arbitrio, 1.5.7.

27. Oppenheimer v. Cattermole [1976] A.C. 249, 278, per Lord Cross.

28. Jilani v. Government of Punjab, Pak. L.D. (1972), S.C. 139, 261, per Sajjad Admad J.

29. Essays in Jurisprudence and Ethics, p. 51.

30. Congreve v. Home Office [1976] Q.B. 629.

31. No one shall be judge in his own dispute; both sides must be heard.

32. Bundesverfassungsgericht, 51; see also 3 B. Verf. G, 231.

33. Mir Hassan v. The State, Pak. L.D. (1969) Lah. 786, 808, per Mushtaq Hussain J. See also Jilani (Supra), at pp. 182, 235, 258, 264, 267.

34. Dialogue between a Philisopher and a Student of the Common Laws of England, s.v. "punishments".

35. The American Supreme Court in the 1930's repeatedly struck down President Roosevelt's "New Deal" legislation because the judges were opposed to its political philosophy. After their death or retirement more sympathetic judges decided otherwise.

36. Law, Liberty and Morality, p. 51 (his emphasis).

37. The Enforcement of Morals, Chap. 1.

38. Op.cit, following John Stuart Mill's Utilitarianism, Liberty and Representive Government.

39. Coral Leasure Group v. Bernett [1981] I.C.R. 503.

40. Knuller v. D.P.P. [1973] A.C. 435; R. v. Ford [1977] 1 W.L.R. 1083.

41. R. v. Bishop [1975] Q.B. 274.

42. Lochner v. New York, 198 U.S. 45, 76 (1905).

43. Law and Other Things, p 48.

44. Legal Essays and Addresses, p. XXV.

45. Op.cit., p. 31.

46. E.g., Jeremy Bentham and John Austin.

47. Thus, Bentham said that punishment for theft consists of two separate laws: one that you ought not to steal, the other that a specified punishment ought to be inflicted if you are found guilty. Kelsen said that there is only one law: if you steal, you ought to be punished, the first part being only a condition precedent of the second. He, therefore, rolls into one law what for Bentham are two laws.